It is safe to say that you are Interested In Learning The Law Of MANIFESTATION Tips Which Will Transform Your Life? Would you like to learn some law of MANIFESTATION procedures that will enable you to show anything in your life?

This long yet uncovering article will demonstrate to you EXACTLY Best practices to do it. You will be given a portion of the best Law of Attraction tips and strategies that will change your whole recognition regarding this matter. In the event that you take after the guidance here, and truly set it in motion, you WILL get comes about. Period!

What's more, I am not discussing some "abracadabra hocus pocus" thing here, I am discussing REAL systems and tips, that paying little respect to your convictions, convey comes about over and over...

So put on your reasoning cap, make a plunge with a receptive outlook and non-judgmental demeanor, and

PLEASE, read the article until the point when the end, as it is the best way to get the entire picture! I know it's sort of long, however you mustn't stop midway!

The Law of Attraction Tips and Techniques that need YOU to utilize them in your Life

This is the means by which everything begins...

1) Burning Desire

This is the most crucial exhortation that in the event that you don't ace, the law of fascination just can't work in your life. You should have a clear objective, or even better, a want, that is so enormous, so comprehensive, that you'll be energized and terrified from it in the meantime.

You should have a **BURNING DESIRE**, a relentless enthusiasm and need, to have, to do or to be something throughout everyday life.

What is it you genuinely need? Do you need $1 000, or a fantasy house maybe? Possibly you'd like immaculate wellbeing? Shouldn't something be said about your optimal accomplice and perfect partner? What about these things?

Have a Burning Desire

Choose what you need and think about it consistently, until the point when it turns into a genuinely consuming, fixating, hazardous want!

2) Don't stress over the "How" Yet

It isn't your business to know everything toward the begin. On the off chance that you have an objective, it must be concrete, absolutely characterized, and you should see obviously the last picture, don't stress for the subtle elements on the most proficient method to accomplish it yet. The way will be appeared to you.

Try not to misunderstand me, this doesn't imply that the Universe will do it for you! It won't, it will do it THROUGH you, by demonstrating to you the way and the means that you have to take, with a specific end goal to get to where you're going.

So don't disappoint on the off chance that you don't have a clue about the way yet. Remember your objective constantly, trust that you will accomplish it, imagine the last picture, and know profoundly that the way will be appeared to you.

A decent case for this would climb a mountain, which you've never climbed. There are likewise no signs for the mountain track, yet the best is noticeable and you know where you're going.

You simply need to see the best and trust that you will arrive... the way will be normally appeared to you... you can't know every one of the subtle elements, every one of the alternate routes and every one of the ways before you begin climbing right? What's more, regardless of whether you get lost now and again, which will occur every once in a while, you will in the end discover the way, in light of the fact that your objective is perfectly clear in your brain.

Try not to Worry About the HOW Yet

3) Autosuggestions and Affirmations

Autosuggestions are the genuine article. This is the methods by which we impact our subliminal personality, and make every one of our wants show in our lives. Compose capable assertions and read them consistently.

Compose something like "I realize that I can have anything I want and I trust that through the Law of Attraction and the all-powerful Universe, I get everything that I need. I am a champ, and I know it!"

Autosuggestions can persuade your subliminal personality that what they say is extreme truth, which thus will give you heaps of confidence. Make your own particular rundown important to your objectives and dreams, and read them as frequently and to the extent that this would be possible.

Perusing Affirmations and Autosuggestions

This may be the greatest weapon that you can use for the fulfillment you had always wanted! All the immense pioneers in everything without exception know about this effective apparatus, and they are ALL utilizing it into their life. So for what reason not to take after their means?

4) Increase Your Faith

As Mr. Slope stated, Faith is for sure, the head physicist of the psyche. At the point when your musings are joined with confidence, your intuitive personality takes these vibrations and it right away interprets them in their profound partner by transmitting them straightforwardly to the Universal Subconscious Mind.

There is just a single method to grow **REAL Faith**, not simply inspiration that will vanish the precise following day. This technique can be summed up in one sentence: "The redundancy of similar contemplations again and again and over again is the thing that prompts expanded confidence. You do this by attestations and by consistent reiteration"

Increment Your Faith

Regardless of whether you rehash a deceive yourself, again and again, your subliminal personality will in the long run begin to trust in those things and they will turn out to be genuine for you! So begin to utilize your confidence appropriately and never again let dread, uncertainty and hesitation enter your psyche!

5) See It First-Visualize

Utilizing perceptions is without a doubt an exceptionally positive and extraordinary path, to prepare and set up your brain to dispose of all the negative idea designs and supplant them all with a fruitful, cheerful, upbeat and prosperous line of reasoning.

There are many types of visualizations for this purpose. One of them includes:

Blank Canvas: Clear your mind of all the thoughts and then picture/visualize a large blank Slowly, and little by little, draw the things that you want to bring into your life on that canvas. It is a good idea to pick the one thing you would like to have, be or do in your life, and every time while using the blank canvas, imagine the same image.

See Yourself in it: Visualize yourself living with the thing you want (e.g. a great job, a new business, achieving perfect health, a wonderful family). Create imaginations of everything you will be doing when you are already living that dream, and soon you will realize that the more you do this, the higher and more believable the chances of realizing that dream.

Action: Don't just imagine static pictures, but imagine the actions! Try to feel the movements, hear the words you'd be saying, hear how others congratulate you on your success.

6) Have Boundless Gratitude

Aligning yourself with positive energy of gratitude will attract more positive energy to you. The people close to you (mostly family, friends) care about you even if they do not show it outwardly.

Focusing on giving thanks to these people is more fulfilling than doing other unpleasant things like arguing, picking up flaws, or having hidden agendas (being untrue).

Building an attitude of gratitude is not hard. You can easily do this by getting a gratitude journal where you record all the things or people that you are grateful to, and all the things that you are grateful for even if you do not have them yet.

Have Unlimited Gratitude

Keep that journal next to your bed, and remember to write it every day before you sleep. Then read it right before sleeping and reread it every time you wake up in the morning. This will remind you of all the positive things in your life throughout the day, and will keep you in a phenomenal vibration that will attract good and positive things in your life.

7) Envision the Essence
Almost every one of us knows what we want but only a few people know what they TRULY want in life. Start paying attention to your deepest passions and desires and not just the mere wishing in your conscious mind.

Concentrate for a moment on the essence and the big picture. Envision how success will make you feel, have a vision of what the freedom of that success will bring you, the people you will be able to help and all other things that you will be able to do with that success.

Envision the Essence and Get the Whole Picture

Do this regardless of the type of goal you have set in front of you.

Focus more on how the attainment of your goal will make you feel, rather than on all the separate pictures and details. These are extremely important and you will develop them later on, but you must start with the Essence, the Big Picture. See it, Believe it and Do it!

8) Let Your Vision Lead You

Many people often focus their attention on their daily surroundings, pain and struggles among other problems. One trick for the law of attraction is to allow your vision for your life to be the light that will guide your steps in darkness.

The only person who can create a vision (a path for your life) is you, and this vision can be anything that you desire. You can use a vision board (a collage, a poster etc.) to set these great forces in motion and boost your vibration every time you look at them.

Create a Vision Board

So put a vision board in your room, set a motivational wallpaper on your computer and cell phone background, put a picture of your goal on your fridge and wardrobe, and make sure that you remember the feeling the attainment of your goal.

Every time you look at these pictures, that feeling will get reinforced and strengthened which in turn will make your vibration a powerful magnetic force for your dream!

9) Seek Security within Yourself First

Think for a second about this saying from the bible: "To him who has shall be given, and he shall have abundance; but from him who does not have, even that which he has shall be taken away"

Now at first glance this might seem like a terrible and very unfair statement. However, this is an ultimate truth; you just need to understand it... If you think of health, happiness, money, success and so on, as something that we get from life and from the world, this would seem like a very unfair statement...

If people are poor they will become even poorer, if a person is not happy, he or she will become even unhappier, if a person is not healthy, his or her health will deteriorate even more...

However, that's not what Jesus had in mind when he said this... The kind of abundance that he is talking about is the abundance in your inner world, in your thoughts, emotions and internal pictures.

This way, it is very fair... if we think and feel positive

outcomes, we will have more of the same, if we have a poor attitude, we will attract even worse things and we will lose our current belongings as well.

Feel Confident and Seek Security Within

So always seek security from within yourself first! Never try to take something from the world, instead, give! Control your inner world, and create a perfect self-esteem and self-image of yourself. Find security within yourself, and you will always have it manifested in your outer world. That is a law!

of wellbeing, bliss, cash, achievement et cetera, as something that we get from life and from the world, this would appear like an exceptionally out of line explanation...

In the event that individuals are poor they will turn out to be considerably poorer, if a man isn't upbeat, he or she will turn out to be much unhappier, if a man isn't solid, his or her wellbeing will weaken significantly more...

Be that as it may, that is not what Jesus had as a

top priority when he said this... The sort of plenitude that he is discussing is the wealth in your inward world, in your musings, feelings and interior pictures.

Along these lines, it is reasonable... on the off chance that we think and feel positive results, we will have business as usual, in the event that we have a poor disposition, we will draw in much more terrible things and we will lose our present possessions too.

Feel Confident and Seek Security Within

So dependably look for security from inside yourself first! Never attempt to take something from the world, rather, give! Control your internal world, and make an immaculate confidence and mental self view of yourself. Discover security inside yourself, and you will dependably have it showed in your external world. That is a law!

10) Self-Love and Respect

Next on the rundown is to, love yourself. To be verifiable, the law of fascination won't be satisfied on the off chance that you, the subject, don't care for yourself. On the off chance that you can love yourself and love each move of your life, at that point others will do likewise. This is on account of like terms will dependably pull in each other.

This is a standout amongst the most vital standards on how the law of fascination functions. The law of fascination will dependably remunerate any feeling or feeling you are going through with deference of the proportionate esteem appended to it. On this note, it is exceedingly essential to do everything you can.

Have Self Love and Self Respect

So go before your mirror and shout so anyone can hear "I Love Myself!"! Look at yourself straight without flinching. Kiss your hand and genuinely feel love and resemblance for your spirit and body.

You can likewise utilize treatment or other valuable techniques to help fulfill your desire. Be that as it may, your want, your choice and your self discipline is all you have to prevail in this undertaking!

11) You Are What You Think

It is imperative to comprehend that your psyche resembles a psychological bank. This implies in the event that you have constructive contemplations, your own financial balance of feeling incredible, develops. With antagonistic contemplations despite what might be expected, it pulls back from your own record. This takes away every one of the positives from you, abandoning you with a mind brimming with negative vitality.

You Are What You Think About

So dependably be cautious of the musings that you fill your brain with, every single day, and always remember that eventually, we as a whole move toward becoming what we consider!

12) Watch your words

We frequently tend to think little of the energy of words in our lives... The words are solid electro-attractive waves that exposed a ton of energy in them. Consider that each word keeping in mind the end goal to exist must be gone before by an idea. These musings can be negative or positive, in this manner, making negative or positive vibrations.

The word is simply transmitting that vitality on a physical level. At the point when a word is talked, we pick that recurrence with our ears, electromagnetic wave is in a split second sent to our cerebrum, and it makes pictures streak on the screen of our psyche.

These photos can be certain or negative, which thus influence the vibration we're in and the things we pull in our life.

Watch Your Words and Be Careful About What You Say

So be watchful when utilizing your words, with other individuals, as well as with yourself also. Each word said so anyone can hear, resembles a confirmation that we provide for ourselves and to others.

Attempt to reconsider before talking, think profoundly and talk deliberately. Mean what you say, and say what you mean, yet never let your self image to take control and make your words like creepy slugs.

13) Daily Meditation

Another awesome way that can help the law of fascination in be satisfied is by practicing day by day contemplation. You can rehearse this procedure to figure out how to keep your brain very still.

Anyway, it will bring your center consideration again from negative emotions and musings. This will bring the positive sentiments of adoration into moment enactment.

You can take around 15-20 minutes once a day to inhale and center around the sentiments of adoration, wellbeing, achievement and success. Focus your consideration on your inward breaths and exhalations and subsequent to quieting your musings, begin utilizing your creative energy and pondering upon these feelings.

Practice Daily Meditation

You are certain to get these emotions into positive enactment for progress. The more you consider on affection, on wellbeing and flourishing inside, the more its radiation will be felt in your life. In any case, it will likewise make you to be cherished by others with no pressure connected.

14) Do What Inspires You

It isn't an amazing thing that numerous individuals don't generally comprehend what rouses them since they are stuck in their regular numerous obligations. They don't have room schedule-wise to consider motivation or how it can make them cheerful throughout everyday life.

Spend no less than 15 minutes consistently to think or do anything that could make you glad. This might tune in to your main tune, reconnecting with somebody you have overlooked in addition to other things, practicing in your exercise center, written work your book, observing some awesome moving motion picture or narrative, the distance going out for a stroll in the recreation center and tuning in to the flawless flying creatures singing, or notwithstanding climbing.

Do What Inspires You

15) Study the Laws of the Universe

There are laws in the universe, which are correct

laws, similar to the one for gravity. You realize that it works, and you never resist this law. You realize that on the off chance that you bounce from the fifth floor of a building, this won't wind up bravo, so you basically don't do it...

Well for what reason not to realize alternate laws of the Universe, which are similarly vital, perhaps more so than the law of gravity?

These laws can have such an incredible effect of your life, that your new mindfulness will truly change the base of your identity. You will think, feel, act and carry on various...

Concentrate The Laws Of The Universe

These Laws will open new entryways for you, they will demonstrate you more openings, they will give you genuine confidence and expectation since they are EXACT, and you just realize that in the event that you tail them, you will get what you need.

The Law of Attraction is one of these laws, however you know, there are different laws, which are straightforwardly associated with this one, and in the event that you need to comprehend The Law of Attraction better, you should comprehend these laws too.

There's the law of remuneration, the law of pardoning, the law of forfeit, the law of reasoning et cetera... and on the off chance that you have likely viewed the motion picture "The Secret", you'd realize that none of these were said in it.

The Law of Attraction is effective, yet you will never have the capacity to crush all of its potential on the off chance that you don't

comprehend and hone these other awesome laws.

You know showing works for other individuals, yet you feel just as you're missing brief comment it work for you.

You know manifesting works for other people, but you feel as though you're missing something to make it work for you. The main reason a great many people stall out while showing is they don't have the foggiest idea about each fundamental advance to show and co-make with the universe.

Here are the seven stages to show anything you need — including cash.

Stage 1: Get clear on what you need.

On the off chance that you don't know precisely what you need, you can't really find a way to get it going. To show something, you should recognize what you want.

That implies you should get clear on the specifics, enumerating the highlights as though your sign is customized for you.

To begin with this progression, make a rundown of 20-25 things you need to show. Get particular about what you need and rundown the traits in the positive (otherwise known as abstain from utilizing the word don't).

"I need an auto" may arrive you a 1965 El Camino without a working motor.

"I need an utilized SUV with under 30,000 miles on it that is evaluated under $15,000" streamlines your inquiry a lot.

Side note: As you make your rundown, give yourself

authorization to need what you need today and be available to transforming it tomorrow. Judgment of yourself doesn't enable you to show anything.

Stage 2: Ask the universe.

When you have your rundown, it's a great opportunity to increase your flag to the universe by requesting what you need. At the point when the universe is sure about what you need to show, at that point it can help you. On the off chance that you don't ask, it will in any case attempt to help you, however it surmises with reference to what you really want.

Try not to surrender what you get over to risk, request what you need.

There are a bunches of approaches to ask including petition, reflection, representation, and vision sheets. A simple method to request what you need is to compose a letter to the universe.

Approach the universe for what you need once a day makes your solicitations clearer and clearer.

Stage 3: Work toward your objectives.

Showing is the specialty of co-making with the universe. Moving in the direction of your objectives expands your odds of getting what you need. It's additionally fun.

Record 3 activities you can do today to convey you nearer to your objective. In the event that you don't

recognize what to do, utilize Google to make sense of what moves you can make. It's imaginable somebody has battled with a similar issue and has expounded on it. Give yourself a chance to be propelled by what others have done.

Begin making a move and continue making a move until you've achieved your objective.

Stage 4: Trust the procedure.

As you move in the direction of your objective, it might address if showing really works. You may get demoralized and baffled. On the off chance that you are sitting in the battle and pondering when things will happen you aren't putting stock all the while. When you question indication, you're advising the universe to demonstrate showing doesn't work.

The Law of Attraction obliges by sending you encounters that keep you stuck.

To show, you need to confide simultaneously.

At whatever point you wind up questioning, get yourself and say, "I'm getting closer and nearer to my objectives consistently. The universe has my back and it's marvelous."

Rehash this expression until the point that you trust it.

Stage 5: Receive and recognize what you get.

The universe is continually giving you help, however it's not entirely obvious the signs (particularly when they come in startling ways). When you begin to recognize and get signs from the universe, the universe will give you a greater amount of what you need.

A decent method to begin with this progression is to record the confirmation in a diary by the day's end.

Make a point to incorporate anything that transpired amid the day that moved you somewhat nearer to your objective.

• If you are attempting to escape obligation and your charge card organization got to deal with a more fitting installment design, that is prove.

- It can likewise be an uplifting quote addressing you on Facebook.

Stage 6: Keep Your Vibration High.

As indicated by the Law of Attraction, you pull in what you are conveying. To pull in a greater amount of what you need, you need to raise your vibration. Vibrations resemble minimal radio signs you are persistently conveying to the universe. You should tune your flag to a vibration deserving of accepting it (read: remain positive and appreciative).

You should simply to feel satisfaction.

By burning through 10-15 minutes every day (in any event) accomplishing something that influences you to rest easy, anything from viewing a YouTube video or reflecting, you're guaranteeing your vibration remains high.

Keep in mind, a crappy disposition implies you will be compensated with poo. Remaining positive is the most effortless approach to be sure great things are coming.

Stage 7: Clear your protection.

On the off chance that you haven't yet showed what you need, it's reasonable you are opposing what the universe brings to the table. Questions, torment, lingering, dissatisfactions, tension, feelings of trepidation, second thoughts and feelings of disdain are generally types of protection.

What's more, they are absolutely typical.

When you see struggle coming up, recognize it and remind yourself to breath and unwind. It may sound something like, "I'm disappointed once more. I'm opposing once more. I get it. All I need to do now is breath, unwind, and let it come."

In the event that you are experiencing serious difficulties relinquishing issue, discover somebody to help you through it.

Showing your wants is 100 percent conceivable yet, to do as such, you should utilize ALL the means.

1. Get clear on what you need.

2. Ask the universe for it.

3. Take activity (enable the universe to get it going).

4. Trust the procedure.

5. Acknowledge what is being sent to you en route.

6. Increase your vibration.

7. Clear all protection. Meditation takes you beyond the ego-mind into the silence and stillness of pure consciousness. This is the ideal state in which to

plant your seeds of intention

2. Release Your Intentions and Desires

Once you're established in a state of restful awareness, release your intentions and desires. The best time to plant your intentions is during the period after meditation, while your awareness remains centered in the quiet field of all possibilities. After you set an intention, let it go—simply stop thinking about it. Continue this process for a few minutes after your meditation period each day.

3. Remain Centered in a State of Restful Awareness
Intention is much more powerful when it comes from a place of contentment than if it arises from a sense of lack or need. Stay centered and

refuse to be influenced by other people's doubts or criticisms. Your higher self knows that everything is all right and will be all right, even without knowing the timing or the details of what will happen.

4. Detach from the Outcome
Relinquish your rigid attachment to a specific result and live in the wisdom of uncertainty.

Attachment is based on fear and insecurity, while detachment is based on the unquestioning belief in the power of your true Self. Intend for everything to work out as it should, then let go and allow opportunities and openings to come your way.

5. Let the Universe Handle the Details
Your focused intentions set the infinite organizing

power of the universe in motion. Trust that infinite organizing power to orchestrate the complete fulfillment of your desires. Don't listen to the voice that says that you have to be in charge, that obsessive vigilance is the only way to get anything done. The outcome that you try so hard to force may not be as good for you as the one that comes naturally. You have released your intentions into the fertile ground of pure potentiality, and they will bloom when the season is right.

We are Manifestation machines, yet since we have 50 to 70 **THOUSAND** oblivious musings for every day, of which around 80% are **NEGATIVE** we're not showing ponders!

While it's valid that the initial step to utilizing the

law of fascination is moving to more constructive state of mind, the reason individuals surrender working with LOA is that they trust that in the event that they begin thinking positive contemplations, at that point everything they could ever hope for will show.

The law of fascination isn't an enchantment wand. Since the mind specially outputs and stores negative encounters, we need to deliberately, constantly construct the positive mental muscle. We as a whole have layers and layers of stories, restricting convictions, fears and obstructs that have turned into the inside scene of our brains and can't be changed overnight just by speculation positive contemplations.

Keeping in mind the end goal to wind up an ace at showing with the law of fascination, we need to fix the examples that have been put away in our oblivious and supplant them with positive, engaging examples. As it were, rewire the mind.

It might sound entangled, however it's definitely not.

By actualizing every day positive practices in our lives, we will move and raise our vigorous vibration with the goal that we can show from a position of quiet, enlivened activity yielding speedier outcomes. In the event that you start utilizing devices that will embed engaging and positive contemplations into your psyche, you'll be ready to deliver great encounters and results!

Here are nine propensities you can actualize in your every day life at the present time to begin initiating showing vitality:

1. Note what you center around.

Begin by focusing on what you center around. Do you focus on what's going right, or what's turning out badly? When you're chipping away at showing your fantasies, hindrances and difficulties will emerge, however when you center around what's correct, you turn into a fantastic issue solver, which fabricates certainty and rapidly raises your vivacious vibration. You will travel through deterrents speedier.

2. Keep a stress list.

Since the mind is Velcro for negative encounters, it is regular that we stress to such an extent. It's simply the mind's propensity. Keep a stress list for 2 weeks. The moment you begin to stress record it. This not just helps discharge the overwhelming vitality that regularly keeps us stuck, yet toward the finish of 2 weeks you will see none of the stresses were justified. Your mind will have evidence that stress is a misuse of vitality.

3. Practice diaphragmatic relaxing.

Inhale from the paunch, not the chest. This kind of breathing actuates the parasympathetic sensory system (rest-and-process), which helps create a feeling of unwinding and satisfaction and enables us

to be quiet and clear when making enlivened move.

4. Calm the monkey mind with contemplation.

Contemplation calms the monkey mind, which is normally one-sided toward pessimism. Pondering does NOT mean you quit considering. It just means the hold around your considerations diminishes. Reflection encourages us pull back consideration from unpleasant, negative

examples we've made after some time.

5. Move your body in the way feels great to you.

Negative feelings get put away in our bodies on a cell level. Moving is one approach to discharge pressure and negative vitality. It doesn't need to be extreme; you can move, rehearse yoga or go for a walk.

6. Keep an appreciation diary.

Appreciation is one of the least complex approaches

to raise our vibration. When we perceive our incredible fortune and welcome every one of our endowments, it consequently places us in a "vibe decent" lively vibration.

7. Record your objectives and interface with your "why."

Composing your objectives won't just enable you to get clear, yet will enable you to make propelled activity ventures toward your fantasies. Interfacing with your "why" implies associating with the inclination that accomplishing the objective will give you. When we associate with the "why," we raise our vibration and progress toward becoming magnets for pulling in the general population, conditions, and so forth that will enable us

to accomplish our objective.